Beginner's Guide To Boiler Operation

Mastering The Fundamentals Of Boiler Technology And Maintenance

Copyright@2024

Falric Torrington

Table Of Content

Preface 4

    Introduction To The Book 4

    Purpose And Target Audience 6

Chapter 1: Understanding Boilers 8

    1. Definition And Purpose Of A Boiler 8

    2. Types Of Boilers (Fire-Tube, Water-Tube, Electric) 11

    3. Components Of A Boiler (Burner, Heat Exchanger, Controls, etc.) 15

Chapter 2: Principles Of Boiler Operation 20

    1. How Boilers Generate Steam Or Hot Water . 20

    2. The Combustion Process 24

    3. Heat Transfer Basics 28

    4. Efficiency Factors 33

Chapter 3: Boiler Safety 38

    1. Importance Of Safety In Boiler Operation 38

2. Common Hazards (Pressure Issues, Chemical Exposure, Burns) 43

3. Safety Valves And Other Safety Components 48

4. Emergency Procedures And Best Practices ..53

Chapter 4: Operating A Boiler 59

1. Starting And Stopping Procedures................59

2. Normal Operating Parameters......................65

3. Monitoring Operations (Pressure, Temperature, Water Levels) 70

4. Adjusting Controls To Maintain Efficiency And Safety 75

Chapter 5: Routine Maintenance.........................81

1. Importance Of Regular Maintenance............81

2. Daily, Weekly, And Monthly Maintenance Tasks 86

3. Seasonal Maintenance For Peak Efficiency..91

4. Keeping A Maintenance Log 96

Chapter 6: Troubleshooting Common Issues 100

    1. Identifying And Diagnosing Common Problems 100

    2. Step-By-Step Troubleshooting Guide 105

Chapter 7: Regulatory Compliance 110

    1. Overview Of Relevant Codes And Regulations 110

    2. Documentation And Record-Keeping 114

    3. Environmental Considerations 120

Chapter 8: Advances In Boiler Technology 127

    1. Recent Technological Advancements 127

    2. The Role Of Automation And IoT In Boiler Operation 133

    3. Future Trends In Boiler Design And Operation 139

Appendices 144

    Glossary Of Terms 144

Preface

Introduction To The Book

Welcome to "Beginner's Guide to Boiler Operation," a comprehensive resource designed to introduce you to the essentials of operating and maintaining boilers safely and efficiently. This book is crafted specifically for beginners who are new to the field of boiler operation, whether you're a budding engineer, a maintenance technician, or simply interested in understanding how these critical systems work.

The goal of this guide is to provide a solid foundation in boiler operation, demystifying the complexities and ensuring that you have the knowledge needed to start your journey. Throughout this book, we'll cover everything from the basic components of boilers and how they operate to the crucial safety practices that must be followed to prevent accidents and ensure efficient operation.

We recognize that starting out with boiler operation can be daunting given the potential risks and technical challenges involved. Therefore, this book is structured to guide you step-by-step through each critical aspect, using simple language. We aim to make the information accessible and easy to understand, without compromising on the thoroughness needed for such an important subject.

This guide also serves as a practical reference that you can return to time and again, whether to refresh your knowledge or to troubleshoot common issues.

Purpose And Target Audience

Purpose

The primary purpose of "Beginner's Guide to Boiler Operation" is to equip individuals who are new to the field of boiler technology with the fundamental knowledge and skills needed to operate and maintain boilers safely and efficiently. This book aims to simplify the complex concepts and operations associated with boilers, making them accessible and understandable to novices. By providing clear explanations, detailed illustrations, and step-by-step guidance, we hope to remove the intimidation factor associated with boiler operation and foster a sense of competence and safety in the reader.

Target Audience

This guide is intended for a diverse range of beginners including:

- **Aspiring Engineers**: Students or recent graduates in mechanical or industrial engineering who seek practical knowledge about boiler systems as part of their career development.

- **Technical Personnel**: Technicians and maintenance workers who are new to working with boiler systems and need a comprehensive resource to assist in daily operations.

- **Facility Managers**: Individuals responsible for the operation and maintenance of facilities where boilers are a critical component, such as in hotels, hospitals, or manufacturing plants, who may not have specialized training in boiler technology.

- **Homeowners**: Especially those who use boilers for heating and hot water in regions where such systems are prevalent and who wish to understand more about how their systems work and how to maintain them.

Chapter 1: Understanding Boilers

1. Definition And Purpose Of A Boiler

Definition

A boiler is a closed vessel in which water or other fluid is heated. The heated or vaporized fluid exits the boiler for use in various processes or heating applications. Although the term "boiler" suggests boiling water, boilers are typically used to heat water below its boiling point, and they can also be used to heat other fluids besides water.

Purpose

The primary purpose of a boiler is to transfer heat to water to generate steam or hot water. Depending on their design and application, boilers serve a variety of functions across different industries:

- **Residential Heating**: In homes, boilers provide hot water and steam for heating systems, which is distributed through radiators or underfloor heating systems.

- **Industrial Processes**: In industries, boilers are crucial for manufacturing processes that require steam or hot water, such as in the production of chemicals, food, and textiles.

- **Power Generation**: Boilers are key components in power plants, where they generate steam that drives turbines to produce electricity.

- **Commercial Use**: In commercial settings like hotels, hospitals, and large buildings, boilers ensure a continuous supply of hot water and heating, crucial for daily operations and comfort.

Boilers must be operated and maintained with care to ensure efficiency and safety. The importance of understanding boiler components, operation principles, and maintenance requirements cannot be overstressed, as these are critical to preventing accidents and prolonging the life of the boiler.

2. Types Of Boilers (Fire-Tube, Water-Tube, Electric)

Boilers can be categorized into different types based on their design and the medium they use to transfer heat. The three most common types of boilers that you'll encounter are fire-tube boilers, water-tube boilers, and electric boilers. Understanding these types will help you choose the right boiler for your application and understand how they operate.

1. Fire-Tube Boilers

   - **Description**: In fire-tube boilers, hot gases from a fire pass through one or several tubes running through a sealed container of water. The heat from the gases is transferred through the walls of the tubes by thermal conduction, heating the water and ultimately creating steam.

   - **Advantages**: They are relatively inexpensive and easier to operate and maintain compared to other types. Their compact size makes them suitable for small to medium steam demand.

- **Disadvantages**: Fire-tube boilers have a lower steam production rate and cannot handle too much pressure, making them less suitable for high-power applications.

2. Water-Tube Boilers

- **Description**: In water-tube boilers, water circulates in tubes heated externally by the fire. Fuel is burned inside the furnace, creating hot gas which heats water in the steam-generating tubes.

- **Advantages**: These boilers can generate steam at higher pressures and they are better suited for applications requiring high steam output, such as power plants.

- **Disadvantages**: Water-tube boilers are more complex in design, requiring more rigorous maintenance and skilled operation. They are also generally more expensive to install.

3. Electric Boilers

- **Description**: Electric boilers use electricity to heat water. They pass electricity through heating elements which convert electrical energy into heat; the heat is then transferred to water.

- **Advantages**: They are extremely clean and efficient, as they do not involve burning fuel. These are ideal for smaller spaces where emission regulations are strict or where combustion-based boilers cannot be used.

- **Disadvantages**: The cost of electricity can make operating electric boilers more expensive compared to boilers that burn fuel. They are generally used for smaller load applications due to the high cost of electrical energy.

Each type of boiler has its unique benefits and challenges, and the choice of which boiler to use will depend on specific needs including the required heat output, the available space, operational costs, and environmental considerations.

3. Components Of A Boiler (Burner, Heat Exchanger, Controls, etc.)

Boilers, irrespective of their type, are composed of several key components, each crucial for the efficient and safe operation of the system. Here's a breakdown of the essential parts:

1. Burner

   - **Function**: The burner initiates the combustion process by mixing air with fuel (gas, oil, or coal) at the correct ratio to achieve efficient combustion. It is responsible for providing the necessary heat to the boiler.

   - **Importance**: The efficiency of the boiler often hinges on the performance of the burner, as it dictates the heating ability and the rate of fuel consumption.

2. Heat Exchanger

   - **Function**: The heat exchanger transfers the heat produced by the burner to the water or steam circulating within the boiler. In water-tube boilers, water flows through tubes that are heated externally. In fire-tube boilers, hot gases flow through tubes immersed in water.

- **Importance**: It's crucial for the effective transfer of heat and overall efficiency of the boiler system.

3. Controls

- **Function**: Boiler controls include thermostats, pressure controls, safety valves, and other sensors that monitor and manage the boiler's operation. These controls ensure the boiler operates safely and efficiently by maintaining the correct pressure, temperature, and water levels.

- **Importance**: Proper functioning controls are essential for safe boiler operation, preventing conditions that could lead to accidents or failures.

4. Water Feed System

- **Function**: This system automatically supplies water to the boiler from an external source when it's needed. It includes a feed pump that controls the flow of water into the boiler.

- **Importance**: Ensures the boiler maintains an adequate water level to prevent overheating and damage.

5. Fuel System

- **Function**: The fuel system includes all the components that store and supply fuel to the burner. It may consist of a fuel pump, filters, and a delivery system that maintains the correct fuel flow.

- **Importance**: Critical for continuous boiler operation and efficiency, as any interruption in fuel supply can shut down the system.

6. Safety Valves

- **Function**: Safety valves release pressure automatically if the pressure inside the boiler exceeds a preset level, preventing potential explosions.

- **Importance**: They are a critical safety feature that prevents catastrophic failures due to overpressure.

7. Flue Gas Stack

- **Function**: Also known as a chimney, this component expels exhaust gases from the boiler to the outside atmosphere after the combustion process.

- **Importance**: Helps maintain proper combustion by removing exhaust gases, thus preventing them from settling back into the boiler system.

8. Condensate Return System

- **Function**: In steam systems, this system returns the condensate (water formed when steam cools and condenses) back to the boiler for re-use.

- **Importance**: Increases efficiency by recycling heat and significantly reducing the need for fresh water.

These components work together to ensure the boiler operates efficiently, safely, and reliably. Understanding each part's function and importance helps in better managing and maintaining the boiler system, ultimately extending its operational life and performance.

Chapter 2: Principles Of Boiler Operation

1. How Boilers Generate Steam Or Hot Water

Understanding how boilers convert energy from fuel into steam or hot water is fundamental to operating and maintaining these systems efficiently. Here's a step-by-step overview of the process:

1. Energy Input (Fuel Combustion)

- The process begins with the boiler's burner, where fuel (typically gas, oil, or coal) is ignited in the presence of air. This combustion generates a flame and hot gases.

2. Heat Transfer

- The heat from the combustion gases is transferred to the water through the walls of the boiler's heat exchanger. In a fire-tube boiler, hot gases pass through one or more tubes surrounded by water. In a water-tube boiler, water flows inside the tubes, which are heated externally by the gases.

3. Steam Generation

- As the water in the heat exchanger absorbs heat, its temperature rises. For steam boilers, once the water temperature reaches the boiling point (100°C or 212°F at sea level), it begins to transform into steam. This conversion occurs at a constant temperature but increases the water's volume dramatically, creating a significant rise in pressure within the boiler.

4. Steam Separation

- In the case of steam boilers, the generated steam rises to the top of the boiler and is collected in the steam drum, where any remaining water droplets are separated from the steam. This ensures that the steam exiting the boiler is as dry as possible, optimizing its quality and efficiency for whatever application it is needed.

5. Distribution

- The dry steam is then piped out of the boiler for use in various applications, ranging from heating buildings to driving machinery or generating electricity. In systems designed for hot water production, the heated water is circulated directly to heating systems or industrial equipment.

6. Condensation and Recycling

- For steam systems, after the steam releases its heat energy, it condenses back into water. This condensate can be returned to the boiler to be heated again, forming a closed-loop system that increases efficiency and conserves water.

7. Exhaust Handling

- After transferring their heat, the combustion gases are cooler and less buoyant. These gases are expelled from the system through the flue or stack, helping to draw fresh air into the boiler for new combustion cycles.

The principles of boiler operation hinge on efficient fuel combustion, effective heat transfer, and proper steam or hot water distribution. Each stage of the process must be monitored and controlled to ensure the system operates safely and efficiently. This understanding not only aids in daily operation but also helps in troubleshooting issues that may arise during the boiler's lifespan.

2. The Combustion Process

The combustion process in a boiler is a critical step where fuel is burned to generate heat. This process must be carefully managed to ensure it is efficient and safe. Here's an overview of the basic steps involved in the combustion process within a boiler:

1. Fuel Delivery

- Fuel (natural gas, oil, coal, or biomass) is delivered to the burner. The type of fuel used depends on the boiler's design and the availability and cost of fuels.

2. Air Intake

- Air is also introduced into the burner. The amount of air is crucial; too little air results in incomplete combustion, producing soot and hazardous carbon monoxide, while too much air cools the furnace and reduces efficiency.

3. Mixing of Fuel and Air

- In the burner, the fuel and air are thoroughly mixed at the right ratio, which is crucial for efficient combustion. This mixture ensures that the fuel is burned as completely as possible, maximizing heat output and minimizing emissions.

4. Ignition

- The fuel-air mixture is ignited by an ignition source, which could be a spark or a pilot flame, depending on the system. This starts the combustion process, producing heat.

5. Flame Stabilization

- After ignition, the flame must be stabilized to ensure continuous and efficient burning. Flame stabilization is maintained by the design of the burner and controlled fuel and air input.

6. Heat Release

- The combustion of the fuel releases a significant amount of heat, which is transferred to the water or steam in the boiler's heat exchanger. The efficiency of this heat transfer is crucial for the overall efficiency of the boiler.

7. Exhaust Gas Formation

- Combustion produces exhaust gases, including carbon dioxide, water vapor, and, depending on the combustion efficiency, potentially harmful emissions like nitrogen oxides, sulfur dioxide, and particulate matter. It's essential to monitor and control these emissions to meet environmental standards and improve air quality.

8. Exhaust Gas Expulsion

- The exhaust gases are expelled from the boiler through the flue or chimney. This expulsion also helps draw fresh air into the combustion chamber, aiding the combustion cycle.

The combustion process is at the heart of boiler operation, transforming chemical energy in the fuel into thermal energy that heats water or generates steam. Proper management of the combustion process is critical for achieving high efficiency, ensuring safety, and minimizing environmental impact. It requires careful control of the fuel and air supply, monitoring of the combustion and exhaust, and regular maintenance to keep the burner and related components in optimal condition.

3. Heat Transfer Basics

Understanding the basic principles of heat transfer is essential for grasping how boilers function and achieve the efficient transformation of fuel into usable heat. Heat transfer within a boiler occurs primarily through three fundamental mechanisms: conduction, convection, and radiation. Here's an explanation of each and how they play a role in boiler operation:

1. Conduction

- **Description**: Conduction is the process by which heat energy is transmitted through a substance when there is a temperature gradient (difference). In boilers, this occurs in the metal walls of the heat exchanger where heat moves from the hotter area (heated by the combustion gases) to the cooler area (the water or steam).

- **Importance**: Conduction is crucial in the efficiency of heat exchangers in boilers. The material used and the thickness of the metal are key factors in how effectively heat is conducted.

2. Convection

   - **Description**: Convection occurs when heat is carried away by a moving fluid, which can be either a liquid or a gas. In a boiler, this is seen when hot combustion gases move over the heat exchanger, transferring heat to the water inside through the heat exchanger's walls.

   - **Importance**: Enhancing convection improves heat transfer efficiency. Design considerations like the placement and shape of tubes or baffles in boilers are used to increase the flow of the water or the combustion gases, thus maximizing heat transfer through convection.

3. Radiation

   - **Description**: Radiation refers to the transfer of heat through electromagnetic waves. In a boiler, radiant heat is emitted from the flame, combustion gases, and other hot components within the firebox or furnace.

- Importance: Radiative heat transfer is significant in the initial stages of heat transfer in boilers, particularly around the combustion chamber where the materials absorb heat radiated from the flame and then transmit it to the water or steam through conduction and convection.

Key Factors Affecting Heat Transfer in Boilers:

- Surface Area: Increasing the surface area of the heat exchanger (e.g., by using more tubes or fins) can increase the rate of heat transfer by providing more space for conduction and convection to occur.

- Temperature Difference: The greater the temperature difference between the hot gases and the water, the more effective the heat transfer. This is a critical design consideration in boiler efficiency.

- **Flow Rate**: The rate at which water and combustion gases flow over the heat exchange surfaces affects heat transfer. Faster flow rates can enhance convection but might require more careful management to ensure heat is absorbed effectively.

- **Material Properties**: The thermal conductivity of the materials used in the construction of the heat exchanger impacts how quickly heat is conducted. Materials with high thermal conductivity, such as certain metals, are preferred.

Effective heat transfer is essential for the efficient operation of boilers. By optimizing the factors that affect conduction, convection, and radiation, boiler designs can achieve higher efficiencies, thereby reducing fuel consumption and minimizing emissions. This understanding also assists in troubleshooting and improving boiler operations.

4. Efficiency Factors

The efficiency of a boiler is determined by how well it can convert the energy in its fuel into usable heat and how much of that heat is transferred to the water or steam without substantial losses. Several factors influence the efficiency of a boiler, and understanding these can help in optimizing boiler performance and reducing operational costs. Here are the key efficiency factors for boilers:

1. Fuel Quality

- **Description**: The type and quality of fuel used have a significant impact on boiler efficiency. Higher quality fuels burn more completely and efficiently, producing more heat and fewer emissions.

- **Importance**: Using the appropriate fuel for the specific boiler design can maximize combustion efficiency and reduce waste.

2. Combustion Efficiency

- **Description**: Combustion efficiency refers to how well the boiler converts the energy contained in the fuel into usable heat. It is influenced by how completely the fuel burns and how much heat is lost with the exhaust gases.

- **Importance**: Improving combustion efficiency involves fine-tuning the burner operation, maintaining an optimal air-to-fuel ratio, and ensuring that the combustion chamber is in good condition.

3. Heat Transfer Efficiency

- **Description**: After combustion, the boiler must transfer this heat to the water or steam. Heat transfer efficiency depends on the design of the heat exchanger and the condition of its surfaces.

- **Importance**: Scale, soot, or other deposits on heat exchange surfaces can significantly reduce heat transfer efficiency by acting as insulators. Regular cleaning and maintenance are essential.

4. Thermal Insulation

- **Description**: Boilers are equipped with insulation to prevent heat from escaping into the surrounding environment.

- **Importance**: Proper insulation reduces heat loss and increases the overall efficiency of the boiler. Deteriorated or inadequate insulation can lead to significant energy waste.

5. Boiler Controls and Instrumentation

- **Description**: Modern boilers are equipped with advanced controls that optimize combustion and monitor various parameters like temperature, pressure, and water levels.

- **Importance**: Sophisticated controls can adjust the boiler operation dynamically for varying conditions to maintain optimal efficiency. Regular calibration and maintenance of these controls are crucial.

6. Condensate Return

- **Description**: In steam systems, returning condensate to the boiler reduces the need for make-up water and takes advantage of the heat energy still present in the condensate.

- **Importance**: Maximizing condensate return can significantly improve the energy efficiency of a steam system by reducing the energy needed to heat fresh water.

7. Exhaust Gas Temperature

- **Description**: Lower exhaust gas temperatures indicate that more heat is being transferred to the water or steam and less is being wasted up the chimney.

- **Importance**: Implementing heat recovery systems, such as economizers, can help reclaim some of this heat from the exhaust, further improving boiler efficiency.

8. Maintenance and Operation

- **Description**: Regular maintenance ensures that all parts of the boiler function as designed. Skilled operation can adjust the controls to suit the specific needs of the boiler, based on demand and other operating conditions.

- **Importance**: Proactive maintenance and skilled operation can prevent losses due to inefficiencies and extend the life of the boiler.

By focusing on these factors, boiler efficiency can be maximized, leading to cost savings on fuel, reduced environmental impact, and prolonged equipment life. Efforts to improve boiler efficiency are not only economically beneficial but also contribute to energy conservation and pollution reduction.

Chapter 3: Boiler Safety

1. Importance Of Safety In Boiler Operation

Safety in boiler operation is paramount, not only to protect the equipment and facility but also to safeguard human life. Boilers, by their nature, operate under conditions of high temperature and pressure, which, if not properly managed, can lead to catastrophic failures with severe consequences. Understanding and prioritizing safety is essential for anyone involved in the operation or maintenance of boilers. Here are the key reasons why safety is so critical in boiler operation:

1. High-Pressure and High-Temperature Operation

- **Overview**: Boilers generate steam or hot water by heating water under controlled high-pressure conditions. These high pressures and temperatures can be dangerous if not properly contained and regulated.

- **Implications**: Any failure in the boiler system, such as a rupture or explosion, can cause severe injury, fatalities, and significant property damage.

2. Chemical Exposure

- **Overview**: Boiler operation often involves the use of chemicals for water treatment to prevent corrosion and scale buildup. Handling and exposure to these chemicals must be managed safely.

- **Implications**: Improper handling or accidental leaks of these chemicals can lead to health risks such as burns, poisoning, or other serious injuries.

3. Combustion Risks

- **Overview**: The combustion process can produce dangerous byproducts, including carbon monoxide (CO), nitrogen oxides (NOx), and sulfur dioxide (SO2), which are harmful if released into occupied spaces or the environment.

- **Implications**: Adequate ventilation, proper burner operation, and regular emissions monitoring are essential to prevent poisoning and environmental pollution.

4. Operational Errors

- **Overview**: Operator errors due to lack of training or negligence can lead to improper boiler settings, which may cause unsafe operating conditions.

- **Implications**: Comprehensive training and adherence to operational protocols are necessary to prevent accidents caused by human error.

5. Equipment Malfunction

- **Overview**: Boilers rely on several mechanical and electronic components that can fail due to wear and tear, lack of maintenance, or faulty manufacturing.

- **Implications**: Regular inspections, maintenance, and replacement of worn-out parts are crucial to prevent malfunctions that could lead to unsafe conditions.

6. Regulatory Compliance

   - **Overview**: There are stringent regulations and standards governing boiler operation to ensure safety. These include local, state, and federal codes and guidelines.

   - **Implications**: Compliance with these regulations not only ensures safety but also avoids legal and financial repercussions from non-compliance.

Safety First Approach

A safety-first approach must be an integral part of the culture in any facility that operates boilers. This includes implementing comprehensive safety protocols, regular safety training for all personnel, emergency preparedness practices, and a proactive approach to maintenance and inspections. Such measures dramatically reduce the risk of accidents and ensure that both the facility and its personnel are well-protected.

2. Common Hazards (Pressure Issues, Chemical Exposure, Burns)

Operating boilers safely requires awareness of several potential hazards that can pose risks to operators and the facility. Here are some of the most common hazards associated with boiler operation:

1. Pressure Issues

- **Overview**: Boilers operate under high pressure to generate steam or hot water. Any failure in pressure containment can lead to dangerous incidents.

- **Specific Risks**: Pressure issues in boilers can result from overpressure (exceeding the pressure for which the boiler is designed), underpressure (leading to incomplete combustion), or sudden pressure drops (causing flash steam that can lead to explosions).

- **Prevention**: Regular checks and maintenance of safety valves, pressure gauges, and control systems are essential to manage pressure levels within safe limits. Proper training in pressure management techniques is also crucial for operational safety.

2. Chemical Exposure

- **Overview**: Chemicals used in boilers for treating water and preventing corrosion and scale can be hazardous.

- **Specific Risks**: Handling boiler chemicals without proper safety equipment or procedures can lead to chemical burns, poisoning, and respiratory issues.

- **Prevention**: Using appropriate personal protective equipment (PPE), such as gloves, goggles, and respirators, is essential when handling chemicals. Proper training in chemical handling and spill response procedures is also crucial.

3. Burns

- **Overview**: The high temperatures in boiler operations not only pose a risk of burns from direct contact with hot surfaces or steam but also from hot water and fires.

- **Specific Risks**: Contact with exposed pipes, flues, or the boiler itself can cause severe burns. Steam leaks and bursts can result in scalding injuries. Fires can occur from improper handling of fuel or failures in combustion controls.

- **Prevention**: Insulating hot surfaces and providing barriers to prevent accidental contact are vital. Regular maintenance to check for leaks and ensure that all fittings are secure can prevent steam and hot water burns. Comprehensive training in fire safety and emergency response is also necessary.

4. Explosions

- **Overview**: Boiler explosions are severe hazards that can result from various issues, including pressure problems, uncontrolled combustion, and failure of safety devices.

- **Specific Risks**: Explosions can cause extensive damage to property and significant injury or death to personnel.

- **Prevention**: Regular inspections, adhering to prescribed maintenance schedules, and ensuring that all safety devices are functional are crucial steps to prevent explosions. Proper combustion control and ensuring the boiler is operated within design specifications are also critical.

Safety Protocols and Emergency Response

To mitigate these hazards, facilities must establish robust safety protocols and ensure they are strictly followed. This includes routine safety audits, immediate reporting and addressing of safety issues, and a clear emergency response plan tailored to potential boiler operation incidents. Training and drills should be conducted regularly to ensure that all personnel are prepared to handle emergencies effectively. Additionally, fostering a safety-conscious work culture encourages vigilance and proactive management of risks associated with boiler operations.

3. Safety Valves And Other Safety Components

Safety in boiler operation is heavily reliant on various specialized components designed to prevent accidents and manage risks. Among these, safety valves are crucial, but there are several other components equally vital for ensuring safe operation. Here's an overview of these essential safety devices:

1. Safety Valves

 - **Function**: Safety valves automatically release steam or water if the pressure inside the boiler exceeds a predefined limit, preventing potential explosions.

 - **Importance**: They are the primary safety feature in any boiler system, designed to act as a fail-safe by relieving pressure that could otherwise build up to dangerous levels.

2. Pressure Gauges

- **Function**: These gauges monitor the pressure levels within the boiler, providing a visual indication of the boiler's internal pressure.

- **Importance**: Regular monitoring of pressure gauges helps ensure the boiler operates within safe pressure limits. Any discrepancies can indicate issues that might require immediate attention, such as leaks or failing safety valves.

3. Low-Water Cutoffs

- **Function**: This device automatically shuts down the boiler if the water level falls below a safe minimum, preventing the boiler from firing without sufficient water.

- **Importance**: This is critical for preventing boiler dry-firing, which can cause severe damage to the boiler and create a fire hazard.

4. Water Level Indicators

- **Function**: These indicators allow operators to visually verify the water level inside the boiler to ensure it is within safe operational limits.

- **Importance**: Maintaining the correct water level is crucial for the safe operation of boilers, as both excessively high and low levels can lead to equipment damage and accidents.

5. Thermostats and Temperature Controls

- **Function**: These controls regulate the temperature of the water and steam in the boiler.

- **Importance**: Proper temperature control prevents overheating and ensures efficient boiler operation. Overheating can weaken boiler materials and lead to failures.

6. Flame Safeguards

- **Function**: Flame safeguards monitor the presence of a flame in the burner and automatically shut down the fuel supply if the flame goes out unexpectedly.

   - **Importance**: This prevents the accumulation of unburned fuel, which could explode upon reignition or leakage into surrounding areas.

7. Blowdown Valves

   - **Function**: These valves are used to discharge water containing concentrated impurities, maintaining water quality in the boiler and preventing scale and corrosion.

   - **Importance**: Regular blowdown is essential for both safety and efficiency as it helps maintain water quality and reduces the risk of damage to the boiler.

8. Emergency Shutdown Systems

   - **Function**: These systems enable rapid shutdown of the boiler in emergency situations.

- **Importance**: They provide an immediate means to stop boiler operation if critical parameters are breached or if there is an imminent risk to safety.

Maintenance and Testing

All these safety components require regular maintenance and testing to ensure they are functioning correctly. This often involves routine inspections, scheduled testing under simulated failure conditions, and timely replacement or recalibration of faulty or worn parts. The adherence to these maintenance practices is a key part of a broader safety strategy in boiler operations, ensuring that all systems are always ready to respond appropriately to potential hazards.

4. Emergency Procedures And Best Practices

Establishing effective emergency procedures and adhering to best practices are critical components of a safety-conscious boiler operation environment. These protocols ensure that when an incident occurs, responses are swift, coordinated, and effective in mitigating potential damages and injuries. Here's an overview of essential emergency procedures and best practices:

1. Emergency Shutdown Procedures

   - **Description**: Detailed instructions on how to safely shut down the boiler in various emergency scenarios, such as excessive pressure, loss of water, or failure of safety components.

   - **Best Practice**: Ensure that all boiler operators are thoroughly trained and regularly drilled in these procedures. Clearly labeled and easily accessible shutdown controls are vital.

2. Emergency Response Plan

- **Description**: A plan that outlines the steps to be taken in response to different types of boiler emergencies, including who to contact, how to evacuate the area, and how to handle injuries until emergency services arrive.

- **Best Practice**: Regularly review and update the emergency response plan to reflect any changes in personnel, boiler equipment, or facility layout. Conduct regular emergency drills to ensure all staff are familiar with their roles during an incident.

3. Fire Safety Procedures

- **Description**: Specific protocols for handling fires in or around the boiler, which may include the use of fire extinguishers, fire suppression systems, and evacuation routes.

- **Best Practice**: Keep fire extinguishing and suppression equipment readily available and maintained. Train all staff in basic fire safety and specific actions to take in the event of a boiler fire.

4. Leak and Spill Management

   - **Description**: Procedures for dealing with fuel, water, or chemical leaks and spills, including containment and cleanup methods.

   - **Best Practice**: Have spill containment kits and materials accessible near potential leak points. Train personnel on handling and reporting leaks and spills safely and efficiently.

5. First Aid and Medical Emergency Response

   - **Description**: Protocols for dealing with injuries resulting from boiler operations, such as burns, chemical exposures, or physical trauma.

   - **Best Practice**: Maintain a well-stocked first aid kit accessible within the boiler operation area. Provide basic first aid training to all personnel, with specific training for those handling chemicals.

6. Equipment and System Failures

- **Description**: Steps to diagnose and isolate system failures that could escalate into more severe problems.

- **Best Practice**: Implement a routine inspection and maintenance schedule to identify and address potential failures before they lead to emergencies. Train staff to recognize early signs of equipment malfunction.

7. Communication During Emergencies

- **Description**: A clear communication plan that outlines who communicates what information to whom, including internal staff and external emergency services.

- **Best Practice**: Use multiple communication methods, such as alarms, intercoms, and mobile alerts, to ensure that all staff are aware of the emergency. Regularly test communication systems to ensure they are operational.

8. Documentation and Reporting

- **Description**: Procedures for documenting incidents and reporting them to the necessary internal and external authorities.

- **Best Practice**: Keep logs of all incidents, however minor, and review them regularly to identify patterns or recurring issues. Ensure compliance with local and national reporting regulations.

Continuous Improvement

A key aspect of managing boiler safety is continuous improvement. This involves regularly reviewing and refining emergency procedures based on lessons learned from incidents and drills, changes in regulatory requirements, and technological advancements. Encouraging a culture of safety where all employees feel responsible for not only following safety protocols but also contributing to their improvement is crucial for maintaining a safe boiler operation environment.

Chapter 4: Operating A Boiler

1. Starting And Stopping Procedures

Operating a boiler involves critical procedures for both starting and stopping the equipment safely and efficiently. Proper execution of these procedures ensures the longevity of the boiler and the safety of the operation environment. Here's a detailed overview:

1. Starting Procedures

  - **Pre-Start Inspection:**

    - **Description**: Check all boiler components, such as safety valves, gauges, and controls, to ensure they are in good working condition. Inspect the boiler for any signs of wear or damage.
    - **Importance**: This step is crucial to identify any issues that could prevent the boiler from operating safely and efficiently.

- **Check Water Levels:**

  - **Description**: Ensure that the boiler water level is at the correct level before starting. This prevents potential damage to the boiler components.
  - **Importance**: Operating a boiler with inadequate water levels can lead to severe damage, such as overheating and tube failure.

- **Purge the Boiler**:

  - **Description**: Purge all air from the boiler using the purge valve. This is done to ensure that no combustible gases are present inside the boiler that could cause an explosion upon ignition.
  - **Importance**: Purging helps prevent the risk of explosion by clearing out any combustible mixtures within the boiler.

- **Light the Burner**:

  - **Description**: Start the burner using the manufacturer's recommended procedure. This typically involves igniting the fuel with an automatic ignition system.
  - **Importance**: Proper ignition is necessary to start the heating process and must be done according to specific safety protocols to avoid accidents.

- **Monitor the Boiler:**

  - **Description**: Once the boiler is ignited, closely monitor the boiler's pressure, temperature, and water levels through the control system.
  - **Importance**: Monitoring ensures the boiler operates within safe and efficient operational parameters.

2. Stopping Procedures

- Shutdown Demand:

- **Description**: Decide to shut down the boiler based on operational demand or for scheduled maintenance.
- **Importance**: Planned shutdowns are part of routine maintenance and operational efficiency, helping to extend the boiler's life.

- Turn Off Fuel Supply:

- **Description**: Cut off the fuel supply to the burner, ensuring that combustion ceases in a controlled manner.
- **Importance**: Properly shutting down the fuel supply prevents the accumulation of unused fuel, reducing the risk of fire or explosion.

- **Cool Down:**

  - **Description**: Allow the boiler to cool down gradually. Do not attempt to cool it rapidly unless the design permits.
  - **Importance**: Gradual cooling prevents thermal stress and damage to the boiler components.

- **Drain the Boiler (if required):**

  - **Description**: Depending on the type of boiler and the manufacturer's instructions, drain the boiler to remove any accumulated sediment or to prepare it for inspection.
  - **Importance**: Draining helps remove impurities and sediment that can affect the boiler's efficiency and safety.

- **Post-Shutdown Inspection:**
  - **Description**: Conduct a thorough inspection of the boiler to check for any signs of wear or damage that might have occurred during operation.
  - **Importance**: Identifying and addressing issues post-operation can prevent future failures and maintain the boiler's efficiency.

Best Practices

Following these procedures, maintaining detailed operational logs, and conducting regular training sessions for all boiler operators are best practices that ensure safe and efficient boiler operation. These procedures help in creating a routine that supports preventive maintenance and mitigates risks associated with boiler operation.

2. Normal Operating Parameters

Maintaining a boiler within its normal operating parameters is crucial to ensuring efficient performance, safety, and longevity of the system. Here's an overview of the key parameters that need to be monitored and controlled during normal boiler operation:

1. Pressure

  - **Description**: The pressure at which the boiler operates is one of the most critical parameters. It must be maintained within the limits specified by the manufacturer or regulatory standards.

  - **Importance**: Excessive pressure can lead to boiler damage or explosions, while too low pressure may indicate leaks or inefficiencies in the system.

2. Temperature

  - **Description**: The temperature of the steam or hot water produced by the boiler must be closely monitored. The operating temperature will vary depending on the boiler's purpose and design specifications.

- **Importance**: Maintaining the correct temperature is essential for achieving optimal energy efficiency and preventing overheating, which can degrade boiler components.

3. Water Level

- **Description**: For steam boilers, the water level must be kept within safe operational limits at all times to ensure there is enough water to be converted into steam without exposing heating surfaces to the air and causing them to overheat.

- **Importance**: Low water levels can cause the boiler tubes to overheat and fail, whereas too high water levels may lead to poor steam quality or carryover of water into the steam system.

4. Combustion Efficiency

- **Description**: This refers to how effectively the boiler uses fuel to produce heat. It is determined by several factors, including the air-to-fuel ratio and the condition of the burner.

- **Importance**: Proper combustion efficiency ensures maximum energy use from the fuel, reduces fuel consumption, and minimizes emissions of harmful gases.

5. Flue Gas Analysis

- **Description**: Analyzing the composition of the exhaust gases (flue gases) can provide insights into the combustion efficiency and reveal any incomplete combustion issues.

- **Importance**: Regular flue gas analysis helps in adjusting the combustion process to reduce pollutants like carbon monoxide and nitrogen oxides, while ensuring optimal fuel usage.

6. pH Level of Boiler Water

- **Description**: The pH level of the water in a boiler should be maintained within a specific range to prevent corrosion and scaling.

- **Importance**: Proper pH balance helps in extending the life of the boiler by preventing corrosion and scale build-up, which can reduce heat transfer efficiency and increase operational costs.

7. Conductivity

- **Description**: The conductivity of boiler water is monitored to assess the level of dissolved solids and salts.

- **Importance**: High conductivity indicates high levels of dissolved minerals which can lead to scaling and inefficiency. Regular blowdown procedures are required to manage conductivity levels.

Monitoring and Control Systems

Modern boilers are often equipped with sophisticated control systems that continuously monitor these parameters and adjust the boiler's operation automatically to maintain optimal conditions. These systems typically include sensors, programmable logic controllers (PLCs), and user interfaces that display real-time data and alerts.

Operators should be trained to understand these parameters and how to respond if readings fall outside the normal ranges. Regular maintenance and calibration of monitoring equipment also play a crucial role in ensuring accurate readings and the ongoing reliability of the boiler system. By maintaining these normal operating parameters, operators can ensure the boiler operates safely, efficiently, and reliably, thereby extending its service life and reducing the risk of downtime and costly repairs.

3. Monitoring Operations (Pressure, Temperature, Water Levels)

Effective monitoring of boiler operations is crucial to ensure safety, efficiency, and longevity of the system. Here's how to properly monitor the key parameters such as pressure, temperature, and water levels during boiler operation:

1. Monitoring Pressure

- **Tools Used**: Pressure gauges and digital sensors are typically used to monitor boiler pressure. These devices should be strategically placed to provide accurate readings of the steam and water pressure within the system.

- **Procedure**: Check the pressure readings regularly against the boiler's specified operating pressure. It's essential to ensure that pressure relief valves are functional and can relieve excess pressure if needed.

- **Frequency**: Continuous monitoring with alarms set to notify operators if pressure deviates from set operational limits.

2. Monitoring Temperature

- **Tools Used**: Thermometers and thermocouples are common tools for measuring the temperature of the water and steam inside the boiler. Infrared sensors may also be used to detect external surface temperatures as a non-contact method.

- **Procedure**: Regularly verify temperature readings to ensure they are within the range specified for safe and efficient operation. Adjustments may need to be made to the fuel supply or burner operation to maintain optimal temperatures.

- **Frequency**: Continuous real-time monitoring, with alerts configured to inform operators of temperature anomalies.

3. Monitoring Water Levels

- **Tools Used**: Water level indicators and electronic probes are used to monitor the water level within the boiler. High-level and low-level alarms are often part of these systems to prevent water levels from becoming dangerously high or low.

- **Procedure**: Keep a constant watch on water level readings. Low water levels can expose boiler parts to excessive heat, leading to damage, while too high water levels can affect steam quality.

- **Frequency**: This should also be monitored continuously with automatic shutdown features in place to protect the boiler in case of critical water level issues.

Best Practices for Monitoring Boiler Operations

- **Regular Calibration**: Ensure all monitoring instruments are regularly calibrated according to the manufacturer's specifications to maintain accuracy.

- **Routine Inspections**: Apart from automated monitoring, perform manual checks and inspections regularly to confirm the accuracy of the automated systems and identify any potential mechanical issues.

- **Documentation and Logs**: Maintain detailed logs of all monitored parameters. Document any deviations and the corrective actions taken. This will help in troubleshooting, maintenance scheduling, and ensuring compliance with safety regulations.

- **Training**: Operators should be thoroughly trained not only in how to read and interpret the monitoring equipment but also in understanding the importance of these parameters and the implications of any deviations from the norm.

- Integration with Control Systems: Advanced boiler systems integrate monitoring tools with automatic control systems that can adjust the operation of the boiler automatically to maintain optimal conditions. Ensure these systems are correctly configured and functioning as intended.

By diligently monitoring pressure, temperature, and water levels, boiler operators can ensure the unit operates within its optimal range, thereby preventing safety hazards and prolonging the boiler's operational lifespan. This proactive approach to boiler management is essential for maintaining efficiency and preventing downtime.

4. Adjusting Controls To Maintain Efficiency And Safety

Proper control of a boiler system is essential for maintaining operational efficiency and ensuring safety. Adjustments to boiler controls should be made thoughtfully and based on accurate monitoring of the system's performance. Here's how to effectively adjust controls in boiler operation:

1. Control of Fuel Supply

- **Objective**: To manage the rate at which fuel is fed into the combustion chamber to ensure optimal combustion.

- **Adjustments**: Regulate the fuel supply based on heat demand and boiler performance indicators. Modern boilers often use automated fuel feed systems that adjust based on real-time data from sensors measuring oxygen, carbon monoxide, and efficiency levels in the flue gases.

- **Safety and Efficiency**: Proper fuel adjustment prevents under or over-firing, both of which can lead to inefficiency and increased emissions. It also avoids the build-up of unburned fuel, which can be a severe safety hazard.

2. Air-to-Fuel Ratio Controls

- **Objective**: To maintain the correct mixture of air and fuel to achieve complete combustion with minimal excess air.

- **Adjustments**: Use oxygen trim controllers that adjust the air intake based on flue gas analysis. This helps in maintaining the optimal air-to-fuel ratio that promotes complete combustion and reduces excess oxygen that leads to heat loss.

- **Safety and Efficiency**: Ensuring the proper air-to-fuel ratio maximizes fuel efficiency and minimizes the production of harmful emissions like carbon monoxide and nitrogen oxides.

3. Water Level Controls

- **Objective**: To maintain safe water levels within the boiler to prevent damage to the boiler and ensure efficient steam generation.

- **Adjustments**: Automated feedwater systems can adjust the water supplied to the boiler based on the steam demand and the water level readings from electronic probes or float devices.

- **Safety and Efficiency**: Correct water level is critical to prevent boiler dry-firing and water carryover, both of which can lead to boiler inefficiency and increased maintenance costs.

4. Pressure Controls

- **Objective**: To maintain boiler pressure within safe and efficient operational limits.

- **Adjustments**: Pressure controls adjust the firing rate of the boiler and the operation of relief valves. Modulating pressure controls can continuously vary the boiler's firing rate to match the steam load, which is more efficient than on/off or high/low operation.

- **Safety and Efficiency**: Stable pressure levels prevent stress on boiler components and ensure consistent steam quality and energy efficiency.

5. Temperature Controls

- **Objective**: To regulate the temperature of the steam or hot water produced by the boiler.

- **Adjustments**: Temperature controls can modulate the burner or adjust the rate of feedwater entering the boiler. In systems with variable heat demand, controls might include temperature feedback systems that adjust the boiler operation to achieve the desired temperature output.

- **Safety and Efficiency**: Proper temperature control prevents overheating and ensures the energy is used efficiently, lowering operational costs and reducing thermal stress on the boiler.

Best Practices for Adjusting Boiler Controls

- **Continuous Monitoring**: Regularly monitor all critical parameters using sensors and automated systems to ensure timely adjustments are made.

- **Regular Maintenance**: Keep control systems well-maintained, including periodic calibration and replacement of faulty sensors or actuators.

- **Operator Training**: Ensure that all boiler operators are trained in understanding how the control systems work and how to make adjustments safely.

- **Use of Advanced Control Systems**: Employ advanced control systems that can automate adjustments based on comprehensive inputs from multiple sensors, improving both efficiency and safety.

- **Documentation**: Maintain records of all adjustments and settings as part of operational logs to help in troubleshooting and in maintaining operational consistency.

Adjusting boiler controls to maintain efficiency and safety is a dynamic process that requires a good understanding of boiler operations and attentive management. Proper control adjustments not only save energy and reduce costs but also extend the life of the boiler and enhance safety.

Chapter 5: Routine Maintenance

1. Importance Of Regular Maintenance

Regular maintenance of a boiler is crucial for several reasons, all of which contribute to the overall efficiency, safety, and longevity of the boiler system. Here's why consistent maintenance is so important:

1. Safety

- **Overview**: A well-maintained boiler is a safe boiler. Many of the potential hazards associated with boiler operation, such as explosions, fires, and leaks, can be directly attributed to a lack of proper maintenance.

- **Details**: Regular maintenance includes inspections and tests of safety valves, pressure and temperature controls, and emergency shut-off mechanisms. Ensuring these components are in good working order prevents accidents and enhances the safe operation of the boiler.

2. Efficiency

- **Overview**: Boilers that are regularly maintained operate at peak efficiency. Build-ups of soot and scale, poor combustion, and inefficient heat transfer can all reduce a boiler's efficiency, leading to higher fuel costs and increased emissions.

- **Details**: Maintenance tasks such as cleaning firetubes or watertubes, adjusting the burner, and checking the flue for obstructions help to maintain the boiler's efficiency. Ensuring that heat transfer surfaces are clean and that the burner is properly adjusted can significantly impact fuel efficiency and operational costs.

3. Reliability

- **Overview**: Regular maintenance reduces the likelihood of unexpected breakdowns and the associated downtime and repair costs.

- **Details**: Preventive maintenance includes replacing worn parts before they fail, lubricating moving parts, and checking seals and gaskets for integrity. Such practices extend the life of the boiler and ensure it remains reliable over its operational life.

4. Regulatory Compliance

- **Overview**: Adherence to local and national safety standards and regulations often requires documented regular maintenance of boiler systems.

- **Details**: Regular inspections and maintenance are frequently mandated by insurance companies and safety inspectors. Failure to comply can result in fines, voided warranties, and potentially void insurance claims.

5. Cost-Effectiveness

- **Overview**: Maintaining a boiler in good condition helps to minimize the total cost of ownership. The cost of routine maintenance is generally much lower than the cost of major repairs due to negligence.

- **Details**: Routine maintenance helps identify and fix minor issues before they become major problems that require expensive repairs or replacements. This proactive approach can prevent significant financial outlays for emergency situations and reduce the lifecycle costs of the boiler.

Best Practices for Boiler Maintenance

To ensure the benefits of regular maintenance are realized, consider the following best practices:

- **Scheduled Maintenance Plan**: Develop and adhere to a regular maintenance schedule based on the manufacturer's recommendations and industry best practices. This schedule should be adjusted based on the boiler's usage and any unique operating conditions it may experience.

- **Record Keeping**: Maintain detailed records of all inspections, maintenance activities, and repairs. This documentation is crucial for tracking the boiler's history, planning future maintenance, and ensuring compliance with regulatory requirements.

- **Professional Inspections**: While daily checks and basic maintenance can be performed by trained in-house staff, certain inspections and tasks should be conducted by certified professionals. This ensures a higher level of scrutiny and expertise, especially for complex components like electrical systems and safety valves.

- **Training**: Ensure that all personnel involved in boiler operation are properly trained in maintenance procedures as well as safe operation practices. Regular training updates are essential as equipment and standards evolve.

2. Daily, Weekly, And Monthly Maintenance Tasks

Maintaining a boiler involves a series of regular checks and tasks that need to be performed at different intervals. Here is a structured approach to daily, weekly, and monthly maintenance tasks to ensure optimal boiler operation and safety:

Daily Maintenance Tasks

1. Check boiler water level: Ensure the water level is within the recommended operating range. Too high or too low water levels can lead to operational issues and potential damage.

2. Inspect for leaks: Check for water, steam, or fuel leaks which can indicate seal or gasket failures.

3. Monitor boiler pressure: Verify that operating pressure remains within safe limits as per the manufacturer's specifications.

4. Examine the burner operation: Observe the burner flame for stability and color, indications of proper combustion.

5. Verify all safety systems: Test safety valves, pressure gauges, temperature controls, and low-water cutoffs to ensure they are functioning correctly.

6. Check control panel: Review the boiler control panel for proper operation and to ensure no alarms or error messages are displayed.

Weekly Maintenance Tasks

1. Test low-water cutoffs: Manually test the low-water cutoffs to ensure they activate correctly, which is essential for preventing boiler dry firing.

2. Inspect and clean fireside surfaces: Depending on boiler usage, inspect the fireside surfaces for soot or scale buildup, which can significantly affect heat transfer efficiency.

3. Blow down the boiler: Conduct a blowdown procedure to remove sediment and impurities from the boiler water, which helps in preventing scale and maintaining water quality.

4. Check all fittings and connections: Tighten any loose fittings and connections to prevent leaks and maintain system integrity.

5. Examine insulation: Check insulation on pipes and boiler surfaces for any damage or wear to prevent heat loss and energy inefficiency.

Monthly Maintenance Tasks

1. Conduct a thorough inspection of the boiler room: Ensure the boiler room is clean, well-lit, and free of any combustible materials.

2. Clean or replace air filters: Clean or replace air filters on the burner to ensure efficient combustion.

3. Test safety and relief valves: Check operation of safety and relief valves to ensure they open at set pressure levels and do not stick or leak, which is critical for boiler safety.

4. Inspect water treatment systems: Examine water softeners and other water treatment equipment to ensure they are functioning properly, which is vital for preventing corrosion and scale buildup.

5. Calibrate instruments and controls: Regular calibration of instruments and controls ensures accurate operation and response of the boiler system to various inputs.

Additional Considerations

- **Seasonal Adjustments**: Depending on geographic location and seasonal variations, additional maintenance tasks might be needed to prepare for high-demand seasons or to perform an annual system overhaul.

- **Documentation and Compliance**: Keep detailed records of all maintenance activities, as this documentation can be crucial for safety audits, warranty claims, and regulatory compliance.

Regular maintenance is the backbone of safe and efficient boiler operation. By adhering to a detailed maintenance schedule, potential issues can be identified early and rectified before they lead to significant problems, ensuring the boiler operates reliably and efficiently over its lifespan.

3. Seasonal Maintenance For Peak Efficiency

Seasonal maintenance is crucial for ensuring that a boiler operates efficiently throughout the year, adapting to changing weather conditions and operational demands. Here's how to conduct effective seasonal maintenance to keep your boiler system running optimally:

Preparing for Heating Season (Fall)

1. Comprehensive System Inspection: Perform a detailed inspection of the entire boiler system, including the boiler itself, all pipes, vents, and radiators or heat exchangers. Look for signs of wear, leaks, or damage.

2. Cleaning: Clean the boiler's heat exchanger, burners, and other components to remove dirt, soot, and corrosion that could impair efficiency.

3. Efficiency Testing: Conduct efficiency tests to ensure the boiler is operating at optimal levels.

This includes checking for proper combustion and adjusting the air-to-fuel ratio if necessary.

4. System Tune-Up: Replace worn parts, lubricate moving components, and make adjustments to the firing rate, air intake, and other settings to enhance efficiency.

5. Heating System Check: Ensure that all components of the heating system, including pumps, valves, and radiators, are functioning correctly and efficiently.

6. Thermostat Calibration: Calibrate thermostats and other control systems to ensure accurate temperature control throughout the heating season.

Preparing for Cooling Season (Spring)

1. De-scaling: For boilers used in conjunction with cooling systems (like those used for absorption chillers), it's important to descale the boiler to remove any mineral deposits that accumulated during the heating season.

2. Leak Checks: Perform pressure tests and check for leaks, which can be more prevalent after a full season of heating due to expansion and contraction of components.

3. Controls and Sensors Check: Test and recalibrate pressure and temperature sensors and controls, as seasonal shifts can affect their accuracy and responsiveness.

4. Ventilation Check: Ensure that the boiler room has adequate ventilation, especially as temperatures rise, to maintain optimal operating conditions.

5. Fuel System Maintenance: Check fuel lines, filters, and connections for integrity and proper function, cleaning or replacing parts as necessary to prepare for less frequent use during warmer months.

Winter Preparation

1. Insulation Review: Inspect and repair insulation on all exposed pipes and components to prevent heat loss and potential freezing.

2. Emergency Preparations: Ensure all emergency controls and safety devices are in working order, particularly in regions where extreme cold can affect boiler operation.

3. Antifreeze Protection: In extremely cold areas, consider the use of antifreeze additives in the boiler system to prevent freeze-ups in idle equipment.

General Seasonal Tips

- Documentation and Records: Maintain comprehensive records of all inspections, repairs, and adjustments. This documentation is crucial for tracking the performance and maintenance history of the boiler.

- **Staff Training**: Refresh the training of staff on emergency procedures, operational practices, and efficiency tips at the start of each season.

Seasonal maintenance helps adapt the boiler system to the demands of the upcoming season, ensuring it operates efficiently and safely. This proactive approach not only extends the life of the boiler but also optimizes its performance, leading to energy savings and reduced operational costs.

4. Keeping A Maintenance Log

Maintaining a detailed log of all boiler maintenance activities is critical for ensuring efficient operation, complying with regulatory requirements, and facilitating effective troubleshooting and repairs. Here's how to keep an effective maintenance log and why it is important:

Importance of a Maintenance Log

1. Regulatory Compliance: Many regulatory bodies require detailed records of maintenance and repairs as part of compliance checks.

2. Warranty Claims: Accurate logs can support warranty claims by proving that the boiler has been properly maintained.

3. Operational Efficiency: Maintenance logs help identify patterns or recurring issues that may not be obvious, allowing for preventive measures to be taken before major problems occur.

4. Safety: Logs provide a record of safety checks and issues, ensuring that all safety components are regularly inspected and are in good working order.

What to Include in a Maintenance Log

- **Date and Time**: Record when the maintenance activity was performed.

- **Technician's Name**: Include the name of the technician or engineer who performed the maintenance.

- **Description of Task**: Provide a detailed description of what maintenance tasks were performed, including any replacements, repairs, or inspections.

- **Findings and Observations**: Note any issues discovered during the maintenance, even if they were minor and were addressed immediately.

- **Action Taken**: Describe any corrective actions taken, including parts replaced, adjustments made, and any follow-up actions recommended.

- **Operational Data**: Record relevant operational data from before and after the maintenance, such as pressure levels, temperature readings, and efficiency measurements, to track performance changes.

- **Signatures**: Have the maintenance person and a supervisor sign off on the completed work to verify the accuracy of the log.

Best Practices for Keeping Maintenance Logs

1. Consistency: Use a standardized format for all entries to ensure consistency and completeness. This can be a physical logbook or a digital system, depending on the facility's requirements.

2. Accessibility: Keep the maintenance log in a location that is easily accessible to all relevant personnel. For digital logs, ensure they are backed up and that access is controlled and secure.

3. Training: Train all personnel involved in boiler operation and maintenance on how to properly record information in the log. Understanding the importance of accurate logging is crucial for compliance and operational integrity.

4. Review and Audit: Regularly review the maintenance logs as part of operational meetings or safety audits. This can help in early detection of potential issues and in assessing the effectiveness of current maintenance practices.

5. Integration with Other Systems: If possible, integrate the maintenance log with other operational management systems to streamline workflows and enhance data analysis capabilities.

Chapter 6: Troubleshooting Common Issues

1. Identifying And Diagnosing Common Problems

Troubleshooting is a critical skill for anyone responsible for boiler operation. Being able to quickly identify and diagnose common problems helps minimize downtime, prevent extensive damage, and ensure efficient operation. Here are some steps and tips for identifying and diagnosing common boiler issues:

1. Low Steam Pressure

  - **Symptoms**: Insufficient steam pressure can affect the efficiency of boiler operation, leading to inadequate heating or processing capabilities.

  - **Common Causes**: Low water level, steam leaks, faulty pressure gauge, or a malfunctioning burner.

- **Diagnosis**: Check the water level first, then inspect for leaks in the steam lines. Verify gauge accuracy and burner operation.

2. High Steam Pressure

- **Symptoms**: Excessive steam pressure can pose a safety risk, potentially leading to pressure relief valve activation or more severe failures.

- **Common Causes**: Over-firing, faulty pressure controls, or blocked vents.

- **Diagnosis**: Inspect the burner settings and adjust if necessary. Check pressure controls and safety valves for proper operation.

3. Water Leaks

- **Symptoms**: Water leaks can lead to a drop in water level and efficiency, and increase fuel consumption.

- **Common Causes**: Corrosion, faulty gaskets, or loose connections.

- **Diagnosis**: Inspect all visible piping, boiler joints, and gaskets for signs of leakage. Test connections and replace any worn gaskets or seals.

4. Poor Heat Transfer

- **Symptoms**: Inadequate heating, longer time to reach operating temperature.

- **Common Causes**: Scale buildup, soot on heat exchange surfaces, or low water temperature.

- **Diagnosis**: Check for scale on heat exchangers and clean if necessary. Inspect for soot buildup and clean boiler tubes or plates as required.

5. Noisy Operation

- **Symptoms**: Banging, whistling, or other unusual noises during operation.

- **Common Causes**: Air in the system, low water flow, or steam hammer.

- **Diagnosis**: Bleed radiators or steam lines to remove air. Check water flow rate and adjust pumps or valves as necessary.

6. Boiler Not Firing Up

- **Symptoms**: The boiler fails to start or ignite, leading to no heat production.

- **Common Causes**: Faulty ignition systems, lack of fuel, or issues with the thermostat.

- **Diagnosis**: Ensure there is fuel available and that it is reaching the burner. Test ignition components and replace if defective. Check thermostat settings and connections.

Best Practices for Troubleshooting

- **Systematic Approach**: Always begin by checking the simplest potential causes before moving on to more complex ones.

- **Use Manuals and Schematics**: Refer to the boiler's manual and schematics for specific diagnostic procedures and to understand the normal operation parameters.

- **Keep Tools and Instruments Handy**: Use appropriate tools and instruments such as pressure gauges, multimeters, and leak detectors to accurately diagnose issues.

- **Document Findings and Repairs**: Keep a record of all troubleshooting steps, findings, and repairs made. This documentation can be invaluable for diagnosing future problems and for training purposes.

2. Step-By-Step Troubleshooting Guide

Troubleshooting boiler problems effectively requires a systematic approach to diagnose and resolve issues. Below is a general step-by-step guide designed to help boiler operators and maintenance teams address the most common boiler issues:

Step 1: **Observation and Symptom Identification**

- **Action**: Identify the symptoms of the problem (e.g., no heat, low pressure, strange noises).

- **Tools Needed**: Observation skills, boiler operation logs.

- **Tip:** Document all observed symptoms and conditions when the problem occurs to help pinpoint the underlying issue.

Step 2: Consult Documentation

- **Action**: Check the boiler's manual and any operation logs for any similar issues or manufacturer-specific troubleshooting steps.

- **Tools Needed**: Boiler manuals, maintenance logs.

- **Tip:** Pay special attention to any manufacturer warnings or tips, as these can provide valuable shortcuts in troubleshooting.

Step 3: Basic Checks

- **Action**: Perform basic checks such as ensuring there is power to the boiler, checking for obvious visible damage or leaks, and verifying that all switches and valves are in their correct positions.

- **Tools Needed**: Basic hand tools for inspections, flashlight.

- **Tip**: These checks can often resolve simple issues without needing further troubleshooting.

Step 4: Specific Diagnostic Tests

- **Action**: Depending on the symptom, perform specific diagnostic tests.

  - For no heat or firing issues: Check fuel supply, ignition systems, and thermostat settings.

  - For pressure problems: Test pressure gauges for accuracy, inspect safety valves and pressure settings.

  - For leaks: Conduct a pressure test or visually inspect joints, pipes, and valves.

- **Tools Needed**: Multimeter, pressure gauge, leak detection fluid.

- **Tip**: Use appropriate personal protective equipment and follow safety protocols while performing these tests.

Step 5: Component Testing

- **Action**: Isolate and test individual components suspected of causing the issue, such as pumps, burners, or sensors.

- **Tools Needed**: Component-specific tools, replacement parts for testing.

- **Tip**: Replace suspect parts temporarily with known good parts to confirm the diagnosis.

Step 6: Adjustments and Repairs

- **Action**: Make necessary adjustments or replace faulty components as identified in previous steps.

- **Tools Needed**: Wrenches, screwdrivers, replacement parts.

- **Tip**: After adjustments or repairs, double-check all connections and settings to ensure everything is reassembled correctly and safely.

Step 7: Functional Test

- **Action**: Restart the boiler and monitor its operation to ensure that the problem has been resolved and that the boiler is operating according to its specifications.

- **Tools Needed**: Boiler controls, monitoring equipment.

- **Tip**: Continue to monitor the boiler closely for a period to ensure no further issues arise.

Step 8: Documentation

- **Action**: Document the problem, the diagnosis, the solution applied, and any parts replaced.

- **Tools Needed**: Maintenance log, digital or paper recording systems.

- **Tip:** Good documentation helps in future troubleshooting and can be crucial for warranty claims or inspections.

Chapter 7: Regulatory Compliance

1. Overview Of Relevant Codes And Regulations

Compliance with codes and regulations is a critical aspect of boiler operation and maintenance. These standards ensure safety, efficiency, and environmental protection. Here's an overview of the primary regulatory frameworks and standards that govern boiler operations:

1. ASME Boiler and Pressure Vessel Code (BPVC)

  - **Description**: Developed by the American Society of Mechanical Engineers (ASME), this is one of the most widely adopted codes worldwide. It provides comprehensive rules for the design, fabrication, installation, and inspection of boilers and pressure vessels.

  - **Importance**: Compliance ensures that boilers are safe and capable of operating under the pressures and temperatures typical in such systems.

2. National Board Inspection Code (NBIC)

- **Description**: Issued by the National Board of Boiler and Pressure Vessel Inspectors, the NBIC provides standards for the installation, inspection, and repair and/or alteration of boilers, pressure vessels, and pressure relief devices.

- **Importance**: Adhering to the NBIC helps in maintaining the integrity and safety of existing equipment through regular inspections and repairs.

3. National Fire Protection Association (NFPA) Codes

- **NFPA 85**: Boiler and Combustion Systems Hazards Code. This code provides essential safety requirements for the operation of boilers and the management of associated fire hazards.

- **Importance**: Ensures that fire safety measures are in place to protect against hazards related to fuel combustion and storage.

4. Occupational Safety and Health Administration (OSHA) Regulations

- **Description**: OSHA regulations, particularly those concerning general industry (29 CFR 1910), outline standards for boiler room operations, including operator training, safety equipment, and emergency procedures.

- **Importance**: Promotes a safe working environment for all personnel involved in boiler operations.

5. Environmental Protection Agency (EPA) Regulations

- **Description**: The EPA issues regulations that impact boilers primarily on emissions of pollutants, such as the Clean Air Act which governs the emission of pollutants including NOx, SOx, and particulate matter.

- **Importance**: Compliance helps in reducing the environmental impact of boiler operations and avoiding significant fines.

6. State and Local Codes

- **Description**: In addition to federal regulations, many states and localities have their own specific codes that may be more stringent or tailored to local environmental or safety concerns.

- **Importance**: Ensures compliance with more localized requirements, which may address specific risks or conditions not covered by national codes.

Compliance with these codes and regulations not only ensures legal and operational safety but also enhances the efficiency and lifespan of boiler systems.

2. Documentation And Record-Keeping

In boiler operations, maintaining thorough documentation and record-keeping is essential for compliance, safety, and efficient management. These records serve as proof of adherence to regulatory standards, facilitate troubleshooting, and help in the efficient maintenance and management of the system. Here's what to include in your documentation and some best practices for effective record-keeping.

Essential Documentation

1. Installation Records

- Contents: Details of the boiler installation including date, location, equipment specifications, and information on the installation team.

- Purpose: Helps in tracking warranty information and is crucial for inspections and maintenance planning.

2. Maintenance Logs

- Contents: Records of all maintenance activities performed, including routine checks, repairs, parts replaced, and any modifications made to the system.

- Purpose: Essential for tracking the history of the boiler, assessing its condition, and planning future maintenance. Also important for regulatory compliance.

3. Inspection Reports

- Contents: Detailed findings from each inspection, including the date, inspector's name and credentials, and any issues found or recommendations made.

- Purpose: Provides a formal record of the boiler's condition and compliance with safety standards.

4. Operational Logs

- Contents: Daily records of operational parameters such as pressure, temperature, water levels, and fuel consumption.

- Purpose: Helps in monitoring boiler performance and efficiency, identifying potential issues early, and optimizing operation.

5. Training and Certification Records

- Contents: Documentation of all training sessions attended by boiler operators and maintenance staff, including dates, course content, and certifications obtained.

- Purpose: Ensures and proves that staff are properly trained to operate and maintain the boiler safely and effectively.

6. Safety and Emergency Procedures

- Contents: Documented safety protocols and emergency response procedures specific to the boiler system.

- Purpose: Crucial for ensuring all personnel know how to act in case of an emergency, enhancing safety, and meeting regulatory requirements.

7. Compliance Documentation

- Contents: Records of all compliance measures taken, including emissions testing, safety valve testing, and adherence to local, state, and federal regulations.

- Purpose: Provides proof of regulatory compliance, which is necessary for legal and insurance purposes.

Best Practices for Record-Keeping

- **Digital Storage**: Utilize digital systems for storing records to ensure they are secure, easily accessible, and can be backed up regularly. Ensure that digital records are also protected against unauthorized access.

- **Regular Updates**: Update documentation promptly after any maintenance, inspection, or operational change. This ensures that records accurately reflect the current status of the boiler.

- **Accessibility**: Make sure that relevant personnel have easy access to these records for operational or regulatory needs. This includes training new staff and providing documentation during audits.

- **Retention Periods**: Adhere to industry standards and legal requirements for how long to retain records. This typically involves keeping records for several years, depending on the specific regulation.

- **Regular Reviews**: Periodically review the records to ensure completeness and accuracy. This is also an opportunity to check if the record-keeping practices need adjustments or improvements.

Effective documentation and record-keeping are not only regulatory requirements but also best practices that contribute to the safe and efficient operation of boiler systems. By maintaining comprehensive and accurate records, organizations can ensure operational continuity, compliance, and safety in their boiler operations.

3. Environmental Considerations

When managing boiler operations, it's essential to consider the environmental impact of these systems. Boilers can contribute significantly to air and water pollution if not properly managed. Here are key environmental considerations for boiler operation and strategies to mitigate their environmental footprint.

Air Emissions

1. Pollutants

- Types: Boilers can emit various pollutants including nitrogen oxides (NOx), sulfur dioxide (SO_2), carbon monoxide (CO), particulate matter (PM), and volatile organic compounds (VOCs).

- Impact: These emissions can contribute to air quality problems like smog, acid rain, and respiratory issues in humans.

2. Emission Control Techniques

- Low-NOx Burners: Reduce the formation of nitrogen oxides by managing the combustion process.

- Flue Gas Desulfurization (FGD): Reduces sulfur dioxide emissions, commonly from coal-fired boilers.

- Electrostatic Precipitators and Baghouses: Capture particulate matter before it exits the flue.

- Catalytic Converters: Used to reduce CO and VOC emissions by catalyzing a redox reaction that converts harmful gases into harmless gases like water vapor and carbon dioxide.

Water Usage and Treatment

1. Water Consumption

- Importance: Boilers require significant amounts of water for generating steam and cooling, which can strain local water resources.

- Conservation Measures: Implementing condensate return systems to recycle the steam condensate back into the boiler, minimizing the need for fresh water intake.

2. Water Treatment

- Purpose: Prevents scale and corrosion in boiler systems, which can lead to inefficiencies and increased fuel consumption.

- Environmental Impact: Treatment chemicals and blowdown disposal can lead to pollution if not handled properly.

- Management Strategies: Use environmentally friendly water treatment chemicals and manage blowdown water to minimize contaminants before disposal.

Waste Management

1. Boiler Slag and Ash

 - Production: Coal-fired boilers produce ash and slag as byproducts of combustion.

 - Disposal and Recycling: These materials can often be recycled in construction materials or disposed of in environmentally safe ways.

2. Chemical Disposal

 - Types: Includes used water treatment chemicals and cleaning solvents.

 - Safe Disposal Practices: Follow local regulations for the disposal of hazardous materials to prevent soil and water contamination.

Energy Efficiency

1. Boiler Efficiency

- Importance: More efficient boilers use less fuel per unit of energy produced, reducing overall emissions and operating costs.

- Strategies: Regular maintenance, upgrading to more efficient boiler technology, and implementing energy management systems.

2. Heat Recovery

- Heat Recovery Steam Generators (HRSG): Capture waste heat from hot flue gases to produce additional steam, improving the overall energy efficiency of the system.

- Economizers: Preheat incoming feedwater using residual heat from the flue gases, reducing the energy needed to bring water to boiling point.

Regulatory Compliance

- Adherence to Regulations: Ensure compliance with all relevant environmental regulations and standards, which may include limits on emissions and requirements for reporting and pollution control technologies.

- **Continuous Monitoring**: Implement continuous emission monitoring systems (CEMS) to track emissions and ensure compliance with environmental standards.

Best Practices for Environmental Management

- Regular Training: Educate all boiler operators and maintenance staff on the environmental impacts of boiler operations and the importance of compliance with environmental practices.

- Sustainability Initiatives: Encourage the adoption of sustainability initiatives such as switching to cleaner fuels, investing in renewable energy sources, and participating in carbon offset programs.

By considering these environmental factors and implementing appropriate mitigation strategies, boiler operators can significantly reduce the ecological footprint of their operations while promoting sustainability and compliance with regulatory requirements.

Chapter 8: Advances In Boiler Technology

1. Recent Technological Advancements

The boiler industry has seen significant technological advancements aimed at improving efficiency, reducing emissions, and enhancing operational safety. These innovations not only cater to the increasing regulatory demands but also contribute to more sustainable energy practices. Here are some of the key recent technological advancements in boiler technology:

1. Condensing Boilers

- **Description**: Condensing boilers are designed to capture heat from exhaust gases that would otherwise be lost through the flue. By condensing the water vapor in these gases, the boiler recovers latent heat, significantly boosting efficiency.

- **Impact**: These boilers can achieve efficiencies of up to 98%, significantly reducing fuel consumption and emissions.

2. Biomass Boilers

- **Description**: Biomass boilers use organic materials like wood pellets, agricultural waste, and others as fuel. These materials are carbon-neutral, making biomass boilers an eco-friendly alternative to fossil fuel-based systems.

- **Impact**: They help reduce dependency on non-renewable resources and lower carbon emissions, aligning with global efforts to combat climate change.

3. Integration with Renewable Energy

- **Description**: Modern boilers are increasingly being integrated with renewable energy sources, such as solar power and heat pumps, to reduce the consumption of conventional fuels.

- **Impact**: This integration not only reduces the environmental impact but also lowers operational costs over time.

4. Advanced Control Systems

- **Description**: The use of automation and advanced control systems in boilers has significantly improved. These systems utilize real-time data analytics and machine learning to optimize fuel consumption, adjust operational parameters, and predict maintenance needs.

- **Impact**: Enhances overall efficiency, extends the life of the equipment, and reduces downtime through predictive maintenance.

5. High-Efficiency Heat Exchangers

- **Description**: Advances in materials science have led to the development of more efficient heat exchangers that can withstand higher pressures and temperatures, improving the overall thermal efficiency of boilers.

- **Impact**: These innovations allow for smaller, more efficient designs that are both cost-effective and energy-efficient.

6. Low-NOx and Ultra-Low-NOx Technology

- **Description**: New burner technologies and retrofit kits have been developed to reduce the formation of nitrogen oxides (NOx) during the combustion process. These technologies include staged combustion and flue gas recirculation.

- **Impact**: They are crucial for meeting stringent emission regulations in many parts of the world, helping to improve air quality and public health.

7. Internet of Things (IoT) Integration

- **Description**: IoT technology is being incorporated into boiler systems to allow for remote monitoring and control. This includes real-time tracking of system performance, instant alerts for malfunctions, and remote adjustments.

- **Impact**: IoT integration improves the responsiveness of maintenance teams, enhances operational reliability, and can significantly reduce energy use.

Looking Forward

The future of boiler technology continues to evolve with a strong focus on sustainability, efficiency, and reducing environmental impact. Ongoing research and development are likely to bring forward more innovations, such as:

- **Advanced Materials**: Use of new composite materials that offer better heat resistance and durability.

- **Modular Boilers**: Systems that can scale up or down based on real-time demand, improving efficiency and reducing wastage.

- Carbon Capture and Storage (CCS): Integration of CCS technologies in industrial boilers to reduce CO_2 emissions.

These advancements are shaping the future of boiler technology, making systems more efficient, less polluting, and better integrated with other energy systems. As these technologies become more widespread, they are expected to play a pivotal role in meeting global energy demands sustainably and economically.

2. The Role Of Automation And IoT In Boiler Operation

Advancements in automation and the integration of the Internet of Things (IoT) technologies have revolutionized boiler operations, enhancing efficiency, safety, and predictive maintenance capabilities. Here's how automation and IoT are transforming the landscape of boiler management:

Automation in Boiler Operations

1. Enhanced Control Systems

- **Description**: Modern boilers are equipped with automated control systems that manage every aspect of boiler operation, from fuel intake and burner management to water levels and steam pressure control.

- **Impact**: These systems ensure optimal operation by maintaining the precise conditions required for efficiency and safety. They reduce human error and allow for finer control over boiler parameters.

2. Automated Fuel and Air Adjustment

- **Description**: Automation technologies allow for real-time adjustments to the air-to-fuel ratio and combustion process based on current demand and external conditions.

- **Impact**: This precision ensures maximum combustion efficiency, minimizing fuel waste and reducing emissions.

3. Safety Enhancements

- **Description**: Automated safety features include continuous monitoring of critical parameters with instant shutdown capabilities if unsafe conditions are detected.

- **Impact**: This immediate response prevents accidents, enhances worker safety, and protects equipment from damage.

IoT Integration in Boiler Systems

1. Real-Time Data Monitoring

- **Description**: IoT-enabled sensors can monitor various operational parameters such as temperature, pressure, and flow rates in real time. This data is transmitted to a central system for analysis.

- **Impact**: Real-time monitoring allows operators to immediately identify and respond to issues before they lead to failures, ensuring continuous operation.

2. Predictive Maintenance

- **Description**: By analyzing data collected from sensors, IoT systems can predict when a component is likely to fail or when maintenance is needed, rather than relying on fixed schedules.

- **Impact**: Predictive maintenance minimizes downtime and can significantly extend the life of boiler components by addressing wear and tear proactively.

3. Remote Control and Diagnostics

- **Description**: IoT technology enables remote control of boiler operations, allowing operators to adjust settings and respond to alerts from any location via a connected device.

- **Impact**: This capability is particularly valuable for facilities with multiple boilers or those in remote locations, as it allows for centralized management and immediate response to operational issues.

4. Energy Management and Optimization

- Description: Integrated IoT systems can optimize energy consumption based on predictive analytics, which assesses patterns of use and external factors like weather conditions.

- **Impact**: This optimization leads to more efficient energy use, reducing operational costs and environmental impact.

Future Directions

- Integration with Smart Grids: IoT-enabled boilers can interact with smart grids to dynamically adjust operations based on energy demand and availability, promoting the use of renewable energy sources.

- Advanced Analytics and Machine Learning: Leveraging advanced analytics and machine learning algorithms, boilers can continuously improve their efficiency and adapt to changing conditions without human intervention.

The integration of automation and IoT technologies in boiler operations marks a significant leap toward smarter, more efficient, and safer thermal energy management. As these technologies continue to evolve, they will play a crucial role in the digital transformation of industrial and residential heating systems, leading to smarter energy usage and significant environmental benefits.

3. Future Trends In Boiler Design And Operation

The boiler industry continues to evolve, driven by technological innovations, environmental regulations, and changing energy demands. Future trends in boiler design and operation are likely to focus on increased efficiency, sustainability, and integration with advanced technologies. Here are some key developments expected to shape the future of boiler technology:

Increased Efficiency and Modularity

1. Advanced Materials: New materials that can withstand higher temperatures and pressures are being developed. These materials will enable boilers to operate more efficiently by maximizing heat transfer while minimizing energy losses.

2. Modular Boiler Systems: There is a growing trend towards modular boiler systems that can be scaled up or down based on real-time demand.

This flexibility allows for better energy management and improved efficiency during periods of fluctuating demand.

Integration with Renewable Energy

1. Hybrid Systems: Boilers are increasingly being designed to work in conjunction with renewable energy sources. Hybrid systems that combine solar, wind, or geothermal energy with traditional boiler technology can reduce reliance on fossil fuels and lower carbon emissions.

2. Heat Pump Integration: Integrating heat pumps with boiler systems to recover waste heat and use it in heating applications can significantly increase overall system efficiency and reduce environmental impact.

Smart Technology and IoT

1. IoT and Automation: The use of IoT technology in boiler systems is expected to become more

widespread, providing real-time monitoring, predictive maintenance, and remote control capabilities. This will enhance operational efficiency and allow for proactive maintenance strategies.

Environmental Impact Reduction

1. Emission Control Technologies: As environmental regulations become stricter, new emission control technologies will be essential. Technologies such as advanced scrubbers and filters for reducing particulate emissions, and chemical treatments to reduce NOx and SOx levels, will become standard.

2. Carbon Capture and Storage (CCS): CCS technology might be integrated into boiler systems, especially in industrial applications, to capture CO_2 emissions directly from the flue gases and store it underground, significantly reducing greenhouse gas emissions.

Enhanced Safety Features

1. Advanced Safety Controls: Safety technology will continue to evolve, incorporating more sophisticated sensors and automatic shut-off capabilities to prevent accidents before they occur.

2. Improved Diagnostic Tools: Enhanced diagnostic tools will provide operators with more accurate and detailed information on boiler conditions, helping to prevent failures and extend the lifespan of the equipment.

Decentralization of Heating Systems

1. District Heating: There is a growing trend towards decentralized heating solutions such as district heating, where a single central boiler provides heat for multiple buildings. This system is more efficient and can be more easily integrated with renewable energy sources.

2. Personalization and Customization: Advances in technology will allow for more personalized control over heating systems, enabling users to adjust settings for maximum comfort and efficiency.

The future of boiler design and operation is set to be significantly influenced by advances in technology, shifts in energy sources, and increasing environmental consciousness. These trends will lead to smarter, more efficient, and more sustainable heating solutions that align with global efforts to reduce energy consumption and minimize environmental impact.

Appendices

Glossary Of Terms

Here's a glossary of key terms related to boiler operation:

1. Boiler: A closed vessel in which water or other fluid is heated. The heated or vaporized fluid exits the boiler for use in various processes or heating applications.

2. Burner: A device that mixes fuel and air in the correct proportions and maintains the flame used for heating in a boiler.

3. Condensate: Water formed when steam cools and condenses back into liquid form, often recycled back into the boiler system.

4. Condensing Boiler: A type of boiler that uses a heat exchanger to recover additional energy from the hot gases created by the combustion process.

5. Efficiency: A measure of how effectively a boiler uses the energy from the fuel it burns to produce heat or steam.

6. Emissions: Gases and particles released into the environment as a result of combustion in the boiler.

7. Heat Exchanger: A device used to transfer heat from the hot gases produced by combustion to the water or steam in the boiler.

8. HVAC: Stands for Heating, Ventilation, and Air Conditioning. A system used to control the temperature and air quality in a building.

9. Modulation: The ability of a boiler to adjust its firing rate (input) to meet varying heating demands without turning off completely.

10. NOx: Abbreviation for nitrogen oxides, which are polluting gases produced during combustion, especially at high temperatures.

11. Overfire: To supply more air or fuel than is needed for efficient combustion, often leading to excessive heat and wastage.

12. Pressure Relief Valve: A safety device designed to open and release excess pressure from a boiler if the pressure exceeds safe levels.

13. Scale: Mineral deposits that form on the inside of the boiler tubes due to impurities in the water, which can reduce efficiency and increase the risk of failure.

14. Steam Trap: A device that automatically removes condensate and non-condensable gases without letting steam escape.

15. Water Tube Boiler: A type of boiler in which water circulates in tubes heated externally by the fire and combustion gases.

16. Fire Tube Boiler: A type of boiler where the hot gases from a fire pass through one or more tubes through which water is circulating.

17. Blowdown: The process of removing water from a boiler to control the concentrations of impurities during the steam production process.

18. Flue Gas: Gases emitted directly from the boiler's combustion chamber through the flue or chimney.

19. Low-NOx Burner: A type of burner designed to reduce the emission of nitrogen oxides during the combustion process.

20. IoT (Internet of Things): Network of interconnected devices that communicate and exchange data with each other, enhancing boiler control and monitoring capabilities.

www.ingramcontent.com/pod-product-compliance
Lightning Source LLC
Chambersburg PA
CBHW071209240526
45470CB00018B/1650